Space

©Just Pictures 2016

All rights reserved. This book or any portion thereof may not be reproduced or used in any manner whatsoever without the express written permission of the publisher except for the use of brief quotations in a book review or scholarly journal.

ISBN 978-1546602385

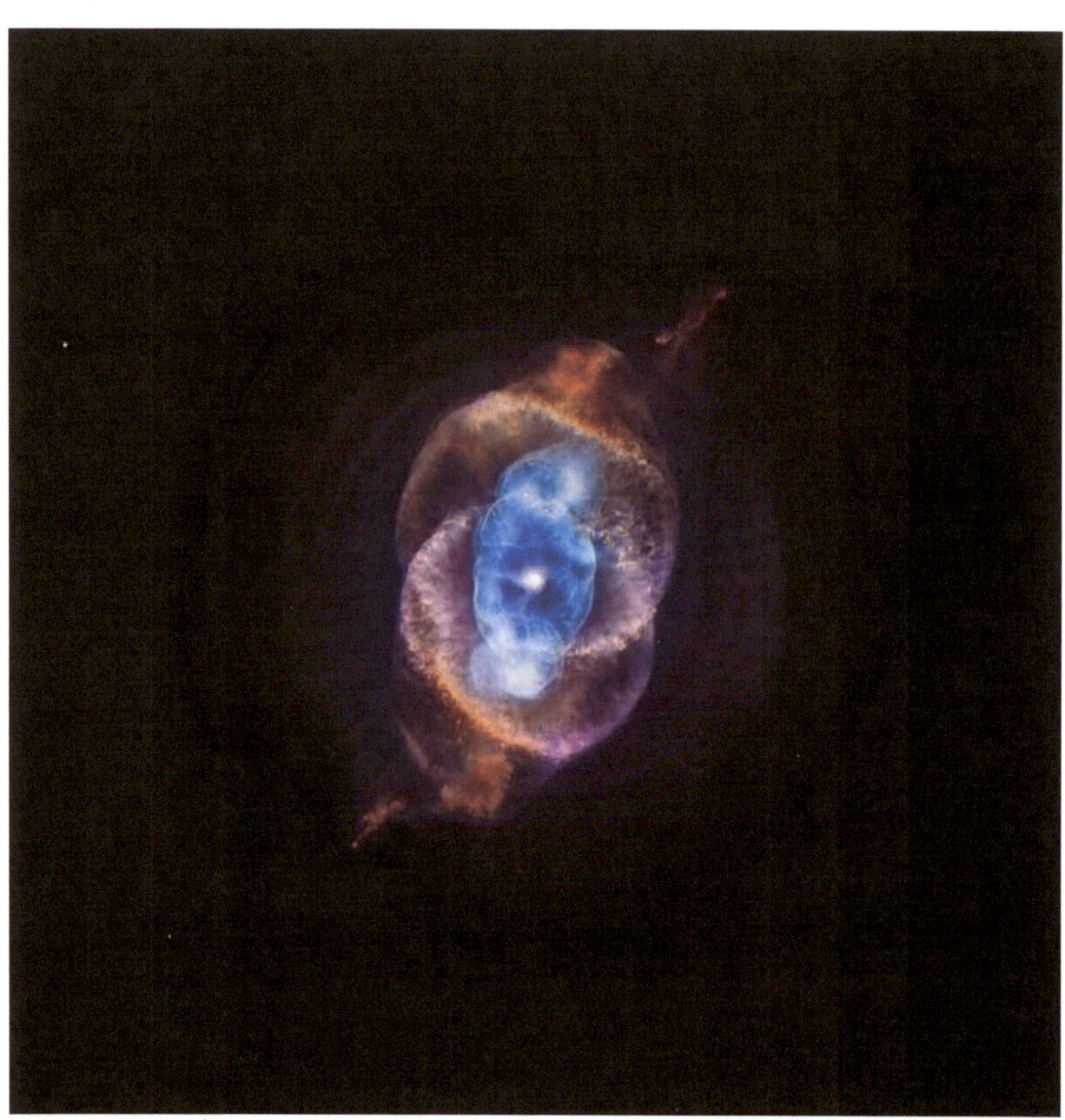

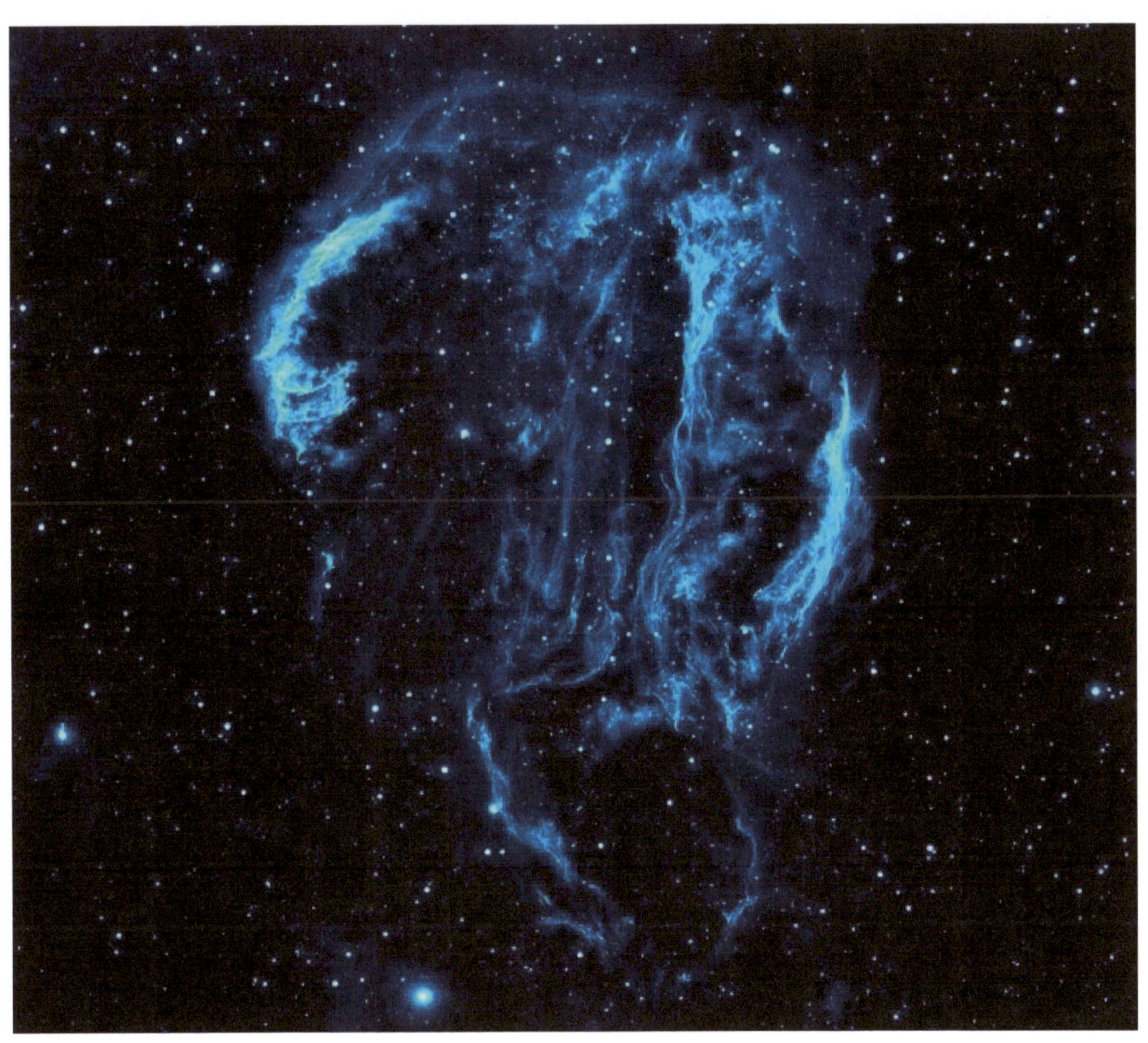

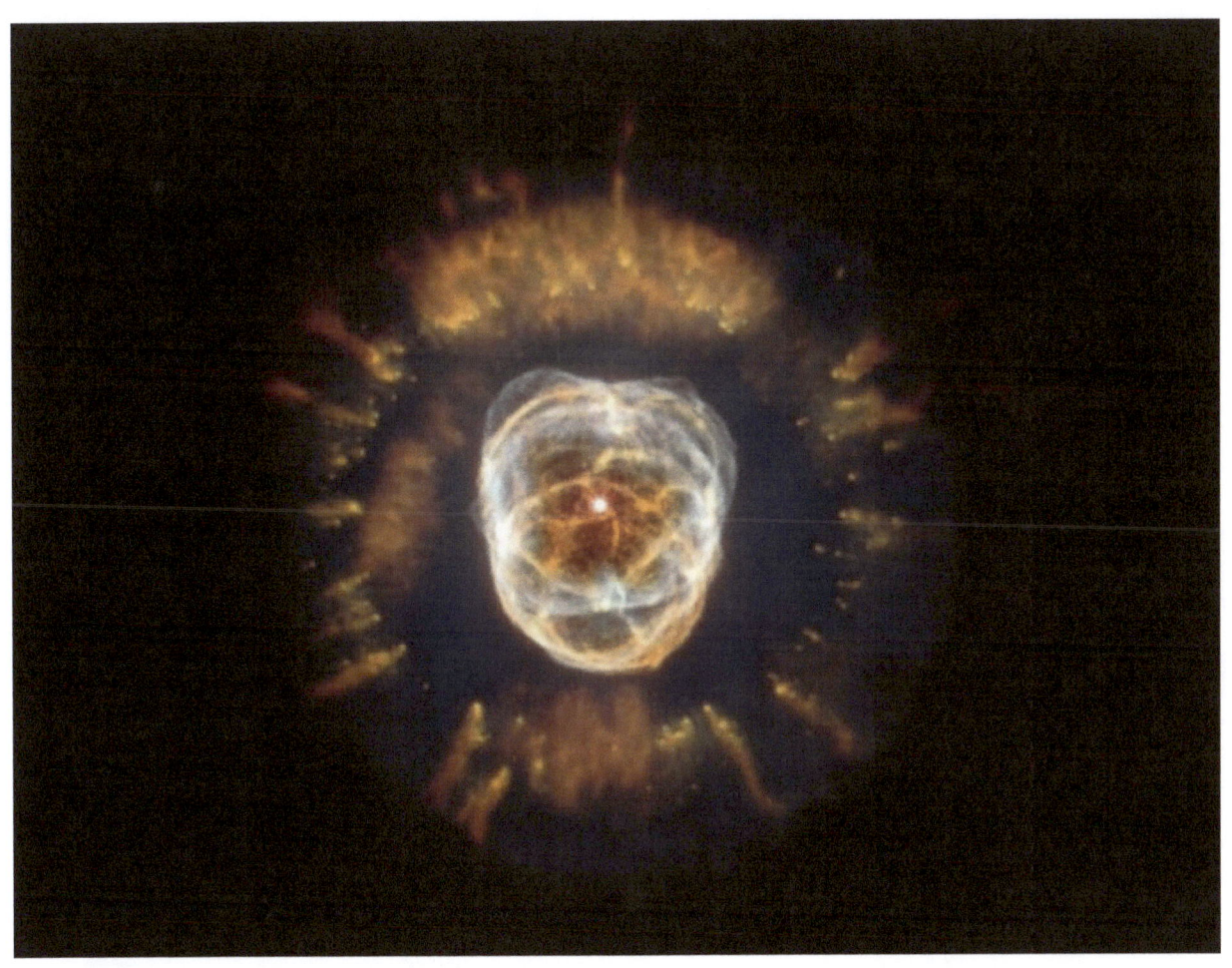

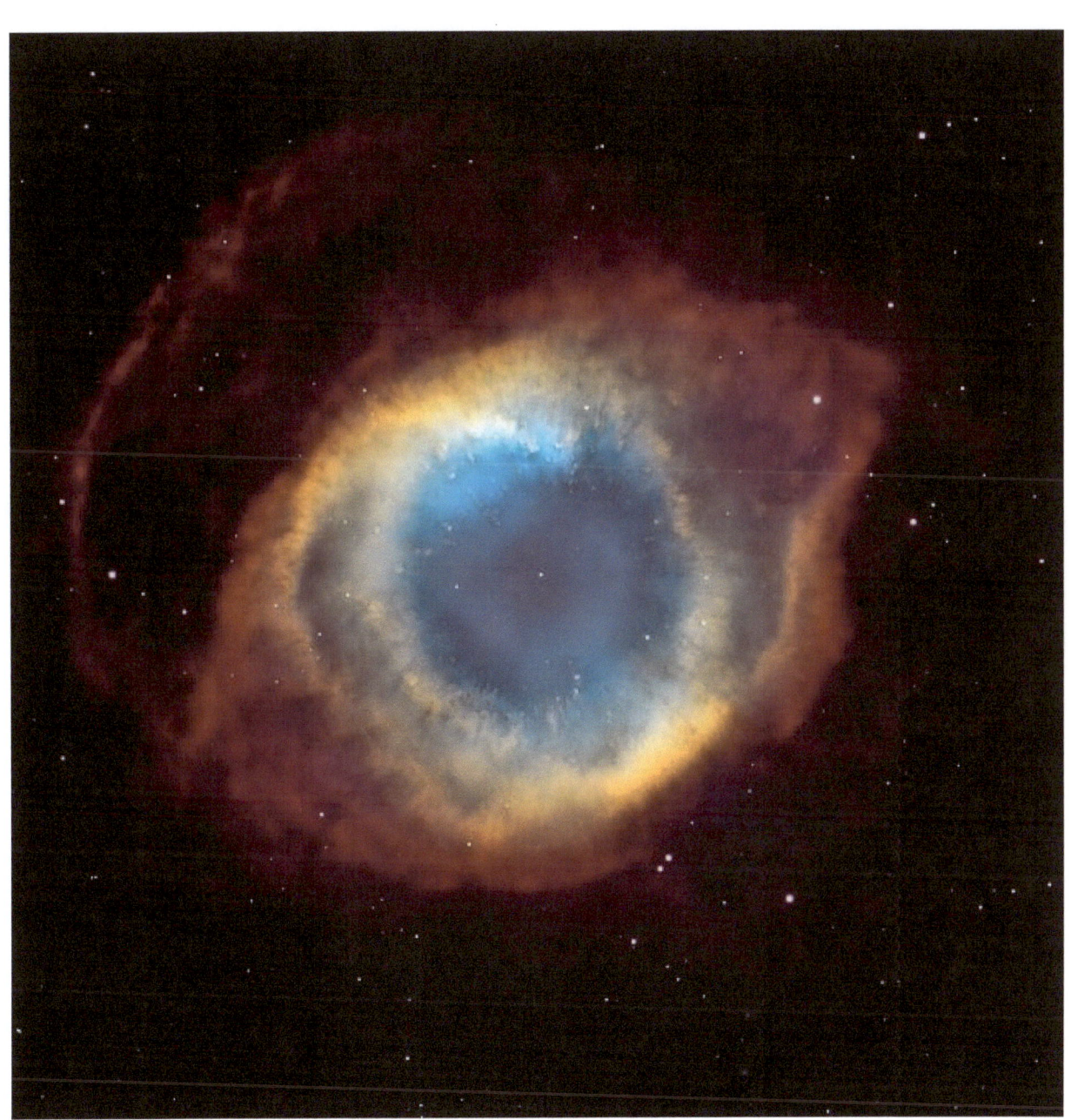

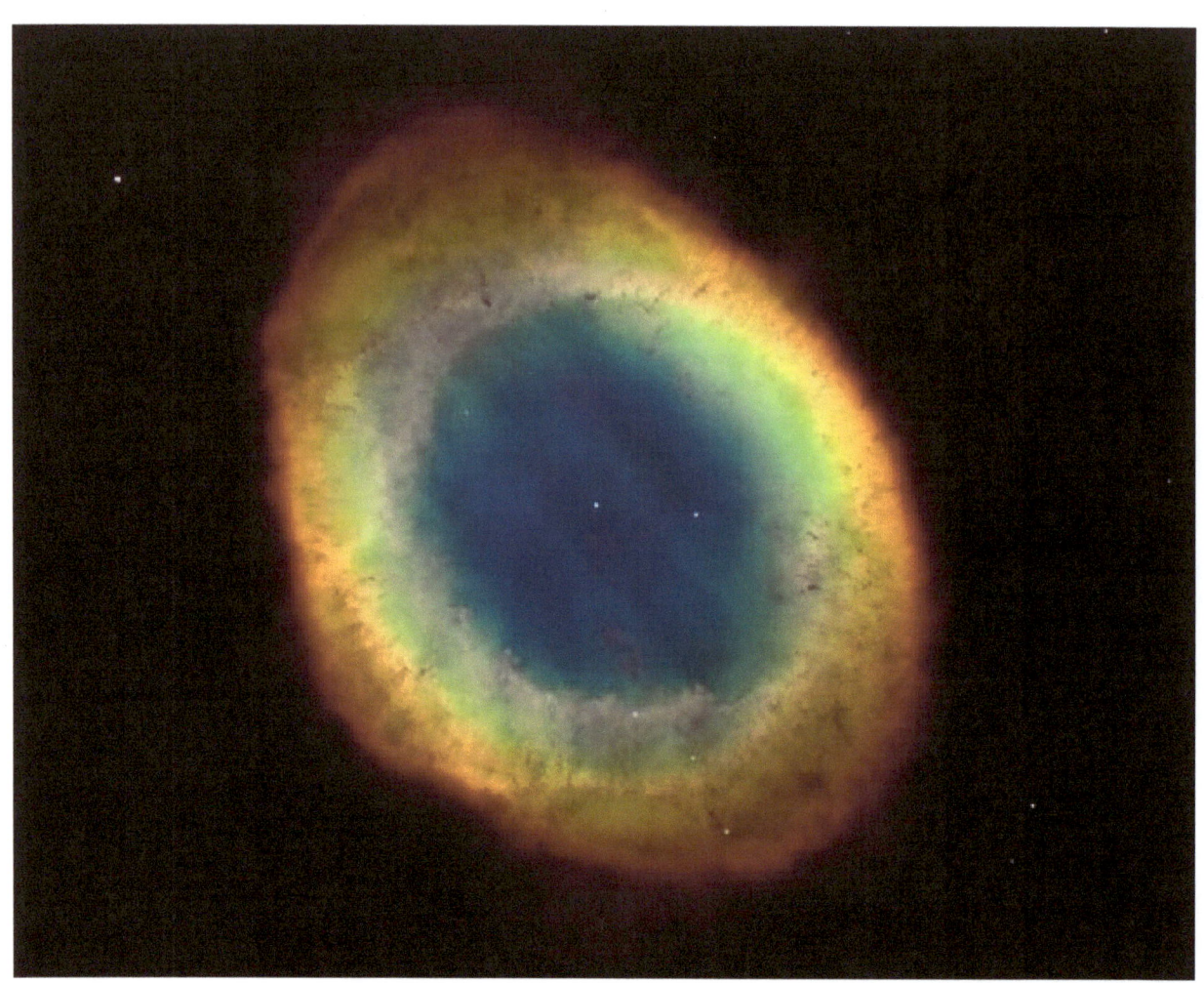

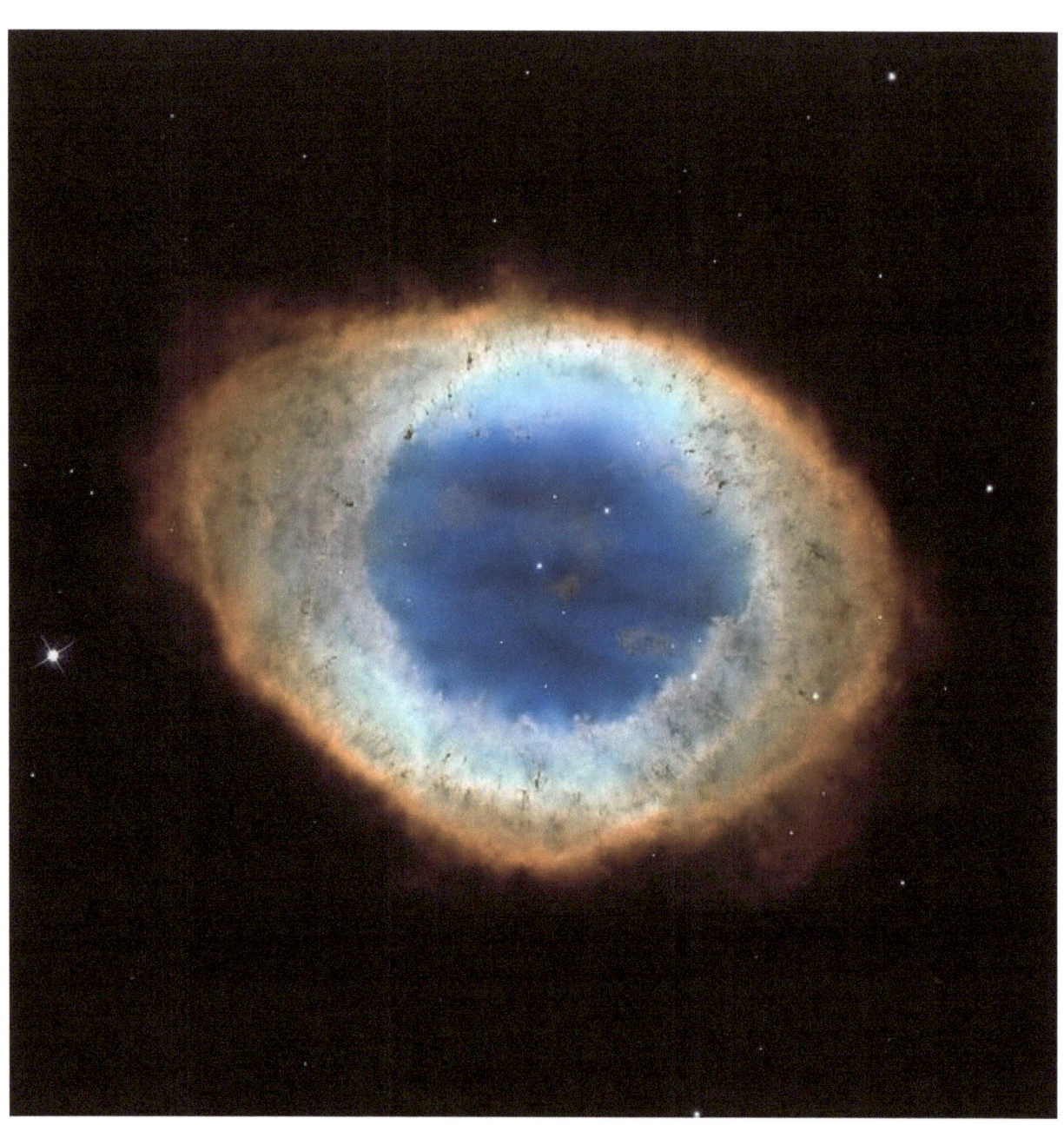

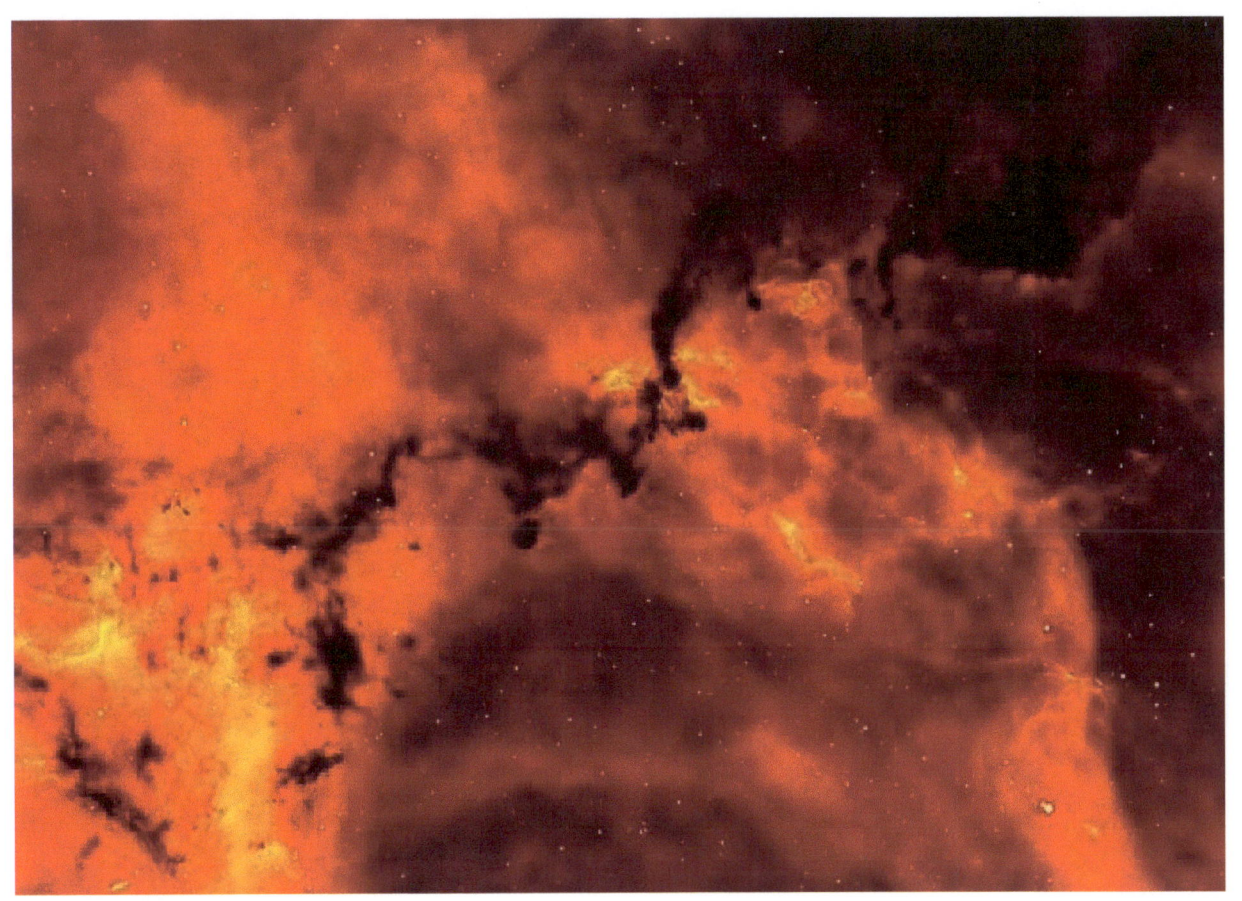